BIBLIOGRAPHIC RECORD TARGET

Graduate Library
University of Michigan

Preservation Office

Storage Number: ____________________

ACV4004
UL FMT B RT a BL m T/C DT 09/12/88 R/DT 09/12/88 CC STAT mm E/L 1
010: : |a 20005885
035/1: : |a (RLIN)MIUG86-B39732
035/2: : |a (CaOTULAS)160649498
040: : |a Univ. of Pennsylvania Libr. |c Miu |d MiU
050/1: : |a QA484 |b .W4
100:1 : |a Weaver, James Henry, |d 1883-
245:00: |a Some extensions of the work of Pappus and Steiner on tangent circles, |c by J.H. Weaver.
260: : |a Lancaster, Pa., |b Press of the New Era Printing Company, |c 1920.
300/1: : |a 1 p. L., 10 p., 1 L, |b diagrs. |c 26 cm.
500/1: : |a Life.
500/2: : |a "Reprinted from the American mathematical monthly, vol.XXVII., January 1920."
502/3: : |a Thesis (PH.D.)--University of Pennsylvania, 1916.
650/1: 0: |a Circle
998: : |c RAS |s 9124

UNIVERSITY OF PENNSYLVANIA

SOME EXTENSIONS OF THE WORK OF PAPPUS AND STEINER ON TANGENT CIRCLES

BY

J. H. WEAVER

A THESIS

PRESENTED TO THE FACULTY OF THE GRADUATE SCHOOL IN PARTIAL FULFILLMENT OF THE REQUIREMENTS FOR THE DEGREE OF DOCTOR OF PHILOSOPHY

PRESS OF
THE NEW ERA PRINTING COMPANY
LANCASTER, PA.

1920

[Reprinted from THE AMERICAN MATHEMATICAL MONTHLY, Vol. XXVII., January 1920.]

SOME EXTENSIONS OF THE WORK OF PAPPUS AND STEINER ON TANGENT CIRCLES.

By J. H. WEAVER, Ohio State University.

Introduction. The figure of three mutually tangent semicircles with their centers in the same straight line was known among the Greeks as the "Shoemaker's Knife" (ἄρβηλος). A few of the properties of the figure are found in the works of Archimedes.[1] Others occur in the Collection of Pappus.[2] After the Greek period we find no work done on the problem until Steiner generalized the results of Pappus and added several others dealing with the perspective properties of the figure.[3] Later Sir Thomas Muir added a theorem giving formulæ for various sets of radii involved.[4] Habicht has discussed some of the properties of elliptic functions connected with the figure[5] while M. G. Fontené has generalized certain formulæ arising from sets of tangent circles.[6]

In the present paper formulæ for the radii of certain sets of circles are developed and used to build up several types of infinite series which may be summed geometrically. Then some general properties of tangents and normals to conics associated with three mutually tangent circles are set forth.[7] These properties lead to a quadrangular-quadrilateral configuration and incidentally furnish some methods for constructing conics. And finally some theorems connected with centers of perspectivity of the various sets of circles are proved.

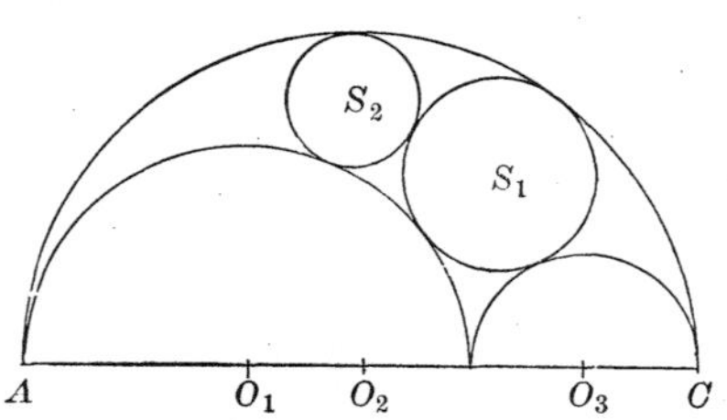

FIG. 1.

1. General Considerations. Let there be two circles tangent internally, with centers O_1 and O_2 and let a circle with center S_n ($n = 1, 2, \cdots$) (Fig. 1) be tangent to these. Then S_n lies on an ellipse with foci O_1 and O_2. If we take the midpoint of O_1O_2 as origin and O_1O_2 as the x-axis, the equation of the ellipse will be

$$\frac{4x^2}{(r_1 + r_2)^2} + \frac{y^2}{r_1 r_2} = 1, \tag{1}$$

where r_1 and r_2 ($r_2 > r_1$) are the radii of the circles (O_1) and (O_2) respectively.

[1] *Works of Archimedes*, ed. Heath, Cambridge, 1897, Lemmas, 4–6.

[2] *Collectio*, ed. Hultsch., Berlin, 1876–8, Vol. I, pp. 209 and ff.

[3] Steiner, *Gesammelte Werke*, Berlin, 1881, Vol. 1, pp. 47–76.

[4] *Proceedings of Edinburgh Math. Soc.*, Vol. 3, p. 119. In the same volume, pages 2–11, J. S. Mackay has collected some of the simpler theorems connected with the problem.

[5] Konrad Habicht, *Die Steinerschen Kreisreihen*, Berne, 1904, 35 pp. In this work are found extensive references bearing on the subject.

[6] "Sur les cercles de Pappus," *Nouvelles Annales de Mathématiques* (4), tome 1918, pp. 383–90.

[7] The center of a circle tangent to two given circles lies on a conic having the centers of the two given circles as foci. This is, of course, equivalent to the definition that the sum or difference of the focal radii is constant. I have called such conics "associated" conics.

[8] In what follows circles will be designated by their centers in brackets.

Let ρ_n be the radius of (S_n) and let the coördinates of the point S_n be x_n and y_n. From a fundamental property of the ellipse we have

$$r_1 + \rho_n = a + ex_n, \tag{2}$$

where $2a = r_1 + r_2$, and

$$e = \frac{r_2 - r_1}{r_2 + r_1}.$$

Pappus has shown that if another circle (S_{n+1}) with radius ρ_{n+1}, center at point x_{n+1}, y_{n+1} and coming after (S_n) in the positive direction around the circles, is tangent to (S_n), the following relation holds

$$\frac{y_n + 2\rho_n}{\rho_n} = \frac{y_{n+1}}{\rho_{n+1}}$$

or

$$\frac{y_n}{\rho_n} = \frac{y_{n-1}}{\rho_{n-1}} + 2 = \frac{y_1}{\rho_1} + 2(n-1). \qquad (3)^1$$

2. Formulæ arising from the figure of three mutually tangent circles with their centers in the same straight line. Let there be three mutually tangent circles (O_1, O_2) and (O_3) having their centers in the same straight line and radii r_1, r_2, and r_3 respectively (Fig. 1). Then let a series of circles (S_n) be drawn tangent to (O_1) and (O_2), the first circle in the series being also tangent to (O_3) and each of the others tangent to the one preceding it in the series. There are two other such sets of tangent circles. The set tangent to (O_2) and (O_3) we will designate as (S_n'), and the set tangent to (O_1) and (O_3) as (S_n''). Let the radii of the various sets be ρ_n, ρ_n' and ρ_n'' respectively, and the coördinates of the centers be x_n, y_n, x_n', y_n' and x_n'', y_n'' respectively. We will now consider the set (S_n).

By means of equations (1), (2) and (3) and the use of induction we have in this particular case, since the y-coördinate of $O_3 = 0$

$$\rho_n = \frac{r_1 r_2 r_3}{n^2 r_3^2 + r_1 r_3 + r_1^2}. \tag{4}$$

This result is arrived at by Muir and Fontené by different methods.[2] Also from equations (2) and (4)

$$x_{n-1} - x_n = \frac{(2n-1) r_3^2 r_1 r_2 (r_1 + r_2)}{[(n-1)^2 r_3^2 + r_1 r_3 + r_1^2][n^2 r_3^2 + r_1 r_3 + r_1^2]} = i_n, \text{ say.} \tag{5}$$

From the geometric properties of the figure

$$\sum_{n=1}^{\infty} \rho_n \tag{6}$$

is a convergent series, and if r_3 approaches the limit 0, then (6) approaches the value $\pi r_2/2$ but is 0 at the limit.

[1] Pappus, *Collectio*, p. 224.
[2] See Introduction.

Also

$$\sum_{n=1}^{\infty} i_n = 2r_1 + r_3.$$

If we define i_n as the nth intercept of the series (S_n) (i_1, projection of O_3S_1 on AC), then (4) and (5) are the formulæ for the nth radius and intercept in the series (S_n). An interchange of r_1 and r_3 will give the corresponding formulæ for the series (S_n'), while an interchange of r_2 and r_3 with r_2 considered negative will give the corresponding formulæ for the set (S_n'').

If $r_2 = 2r_1$ we have the special case

$$\sum_{n=1}^{\infty} \rho_n'' = r_2/2.$$

We will now establish the following theorem.

THEOREM: If two circles (O_1) and (O_2) are tangent internally, and a circle (S) is drawn tangent to these two, such that SO_i $(i = 1, 2)$ is perpendicular to O_1O_2 then it is possible to draw a circle (S') tangent to (O_1), (O_2) and (S) such that the four centers S, S', O_1 and O_2 determine a rectangle.

Proof: Let SO_2 be perpendicular to O_1O_2, and let the coördinates of S and S' be x, y and x', y' respectively and let the radii be ρ and ρ'. Then

$$x = \frac{r_2 - r_1}{2}, \qquad y = \frac{2r_1r_2}{r_1 + r_2}, \qquad \rho = \frac{r_2(r_2 - r_1)}{r_1 + r_2}, \tag{6}$$

and by virtue of equations (1), (2) and (3)

$$y' = y \text{ and } x' = -x,$$

which proves the theorem.

THEOREM: If in the series (S_n), the points S_n, S_{n+1}, O_1 and O_2 determine a rectangle, then $r_1 = nr_3$.

Proof: Equate the values of ρ given in equations (4) and (6).

Let the foot of the perpendicular from S_n to O_1O_2 be P_n.

Let angle $P_nS_nS_{n+1} = \angle B_n$.

Then if $r_1 = kr_3$ (k an integer or a rational fraction)

$$\tan B_n = \frac{(2k+1)(2n+1)}{2(n^2 + n - k - k^2)} \tag{7}$$

and the slopes of the lines of successive centers of the series (S_n) are all rational. Moreover if in (7) k is an integer, $B_k = \pi/2$, and

$$\sum_{n=k+1}^{\infty} (B_{n-1} - B_n) = B_k - \lim_{n=\infty} B_n = \pi/2.$$

From the identity $(\rho_n + \rho_{n+1})^2 = i_{n+1}{}^2 + (y_n - y_{n+1})^2$ we get by using equations (3), (4) and (5) the equation

$$[(2n^2+2n+1)r_3^2+2r_1r_3+2r_1^2]^2 = [(2n^2+2n)r_3^2-2r_1r_3-2r_1^2]^2+[(2n+1)r_3(r_2+r_1)]^2$$

giving a triply infinite set of rational right triangles.

3. Formulæ arising from three mutually tangent circles, one of which is tangent to the line of centers of the other two. Let there be two circles (O_1) and (O_2) (Fig. 2) tangent internally at A and let a series of circles be drawn tangent to these, the first one in the series being tangent to the line O_1O_2 and each of the others tangent to the one preceding it in the series. Let the radius of (O_1) be r_1 and of (O_2) be r_2, and let the centers of the circles be S_n. Then from equations (1), (2) and (3), since $y_1 = \rho_1$ we get by induction

$$\rho_n = \frac{4r_1r_2(r_2-r_1)}{4(n^2-n)(r_2-r_1)^2+(r_2+r_1)^2} \tag{8}$$

$$i_n = \frac{32(n-1)r_1r_2(r_2-r_1)^2(r_2+r_1)}{[4(n^2-n)(r_2-r_1)^2+(r_2+r_1)^2][(4n^2-12n+8)(r_2-r_1)^2+(r_2+r_1)^2]} \tag{9}$$

where for i_n $n \geqq 2$.

Here

$$\sum_{n=2}^{\infty} i_n = 2r_2 - i_1{}^{1}$$

and if r_1 approaches r_2

$$\sum_{n=2}^{\infty} \rho_n$$

approaches $\pi r_2/2$.

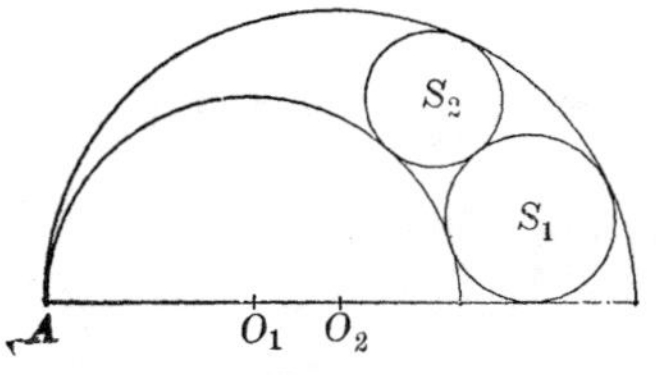

FIG. 2.

THEOREM: If in this series the points S_n, S_{n+1}, O_1 and O_2 determine a rectangle then

$$r_2 = \frac{2n+1}{2n-1}\cdot r_1.$$

Let

$$r_2 = \frac{2k+1}{2k-1}\cdot r_1.$$

Then

$$\tan B_n = \frac{2nk}{n^2-k^2}.$$

Moreover,

$$\sum_{n=k+1}^{\infty}(B_{n-1}-B_n) = B_k - \lim_{n=\infty} B_n = \pi/2.$$

Also the equation $(\rho_n+\rho_{n+1})^2 = (i_{n+1})^2+(y_n-y_{n+1})^2$ gives the triply infinite set of rational right triangles[2]

$$[4n^2(r_2-r_1)^2+(r_2+r_1)^2]^2 = [4n^2(r_2-r_1)^2-(r_2+r_1)^2]^2+[4n(r_2^2-r_1^2)]^2.$$

[1] The distance from the point of contact of S_1 with the diameter AO_2 to the end of this diameter, on the side opposite from the point O_2,, is taken as i_1.—Editor

[2] This result is but a special case of a rational right triangle with sides u^2+v^2, u^2-v^2, and $2uv$.—Editor.

4. Formulæ arising from a series of tangent circles, tangent to two given circles, the first circle in the series being tangent to a line tangent to the smaller circle and perpendicular to the line of centers. Let there be two circles (O_1) and (O_2) tangent internally at A and let $(O_1) < (O_2)$ and let the tangent to (O_1) perpendicular to O_1O_2 be drawn and let a series of tangent circles be drawn tangent to (O_1) and (O_2), the first circle in the series being also tangent to the perpendicular just drawn (Fig. 3).

Then from Pappus, Book IV., lemma XIX, we have

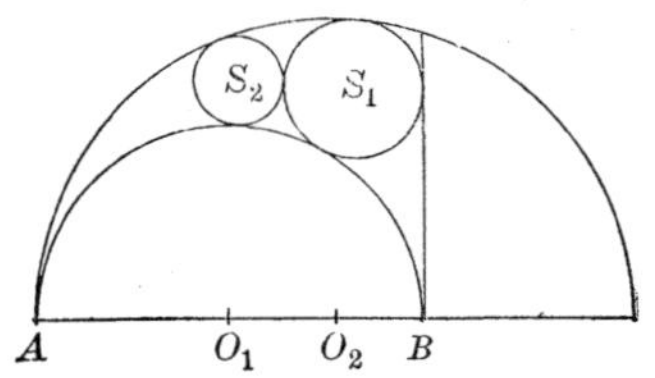

FIG. 3.

$$\frac{r_2 - r_1}{r_2} = \frac{4\rho_1^2}{y_1^2}.$$

Let

$$\frac{r_2 - r_1}{r_2} = m^2,$$

then

$$2\rho_1 = my_1. \tag{10}$$

Using equations (1), (2), (3) and (10) we have by induction

$$\rho_n = \frac{r_1 m^2}{(n-1)^2 m^4 + 2(n-1)m^3 + 1},$$

$$i_n = \frac{r_1 m^3(2 - m^2)[(2n-3)m + 2]}{[(n-2)^2 m^4 + 2(n-2)m^3 + 1][(n-1)^2 m^4 + 2(n-1)m^3 + 1]}.$$

If r_1 approaches r_2, the sum $\sum_{n=1}^{\infty} \rho_n$ approaches $r_2 \cdot \pi/2$ but is zero at the limit. Also $\sum_{n=1}^{\infty} i_n = 2r_1$, if $i_1 = \rho_1$. Let $\angle B_n = \pi/2$. Then since

$$\sin B_n = \frac{i_{n+1}}{\rho_n + \rho_{n+1}}$$

we have the relation

$$n = \frac{1-m}{m^2},$$

and if $n = 1$, $r_1 : r_2 =$ side of decagon inscribed in a circle of unit radius. Here also as in sections (2) and (3) we may obtain an equation

$$[(2n^2 - 2n + 1)m^4 + (4n-2)m^3 + 2]^2$$
$$= [2n^2 - 2n)m^4 + (4n-2)m^3 + 4m^2 - 2]^2 + [m(2-m^2)((2n-1)m + 2)]^2,$$

which gives a doubly infinite set of rational right triangles.

5. Formulæ arising from a series of tangent circles tangent to a given circle and a given straight line. Let (Cf. figure 3) the series of circles (S_n) be tangent to the perpendicular at B and to O_2, (S_1) being also tangent to O_1.

The centers S_n lie on a parabola with vertex O_1 and focus O_2. Using O_1 as origin the equation of the parabola is

$$y^2 = 4(r_2 - r_1)x. \tag{11}$$

Moreover

$$\rho_n = r_1 - x_n \tag{12}$$

and

$$(y_n - y_{n-1})^2 = (\rho_n + \rho_{n-1})^2 - (x_n - x_{n-1})^2. \tag{13}$$

By a substitution from (12) equation (13) reduces to

$$(y_n - y_{n-1})^2 = 4\rho_n\rho_{n-1}. \tag{14}$$

Let $r_1 = \lambda r_2$. Let D_n denote the sum of the odd-numbered terms in the expansion of $(1 + \sqrt{\lambda})^n$: and N_n the sum of the even-numbered terms, that is

$$D_n + N_n = (1 + \sqrt{\lambda})^n,$$

$$D_n - N_n = (1 - \sqrt{\lambda})^n.$$

Then using equations (11), (12) and (14) and induction

$$\rho_n = \left(1 - \frac{N_n^2}{D_n^2}\right) r_1,$$

$$y_n = \frac{2N_n}{D_n}\sqrt{r_1(r_2 - r_1)}. \tag{15}$$

From (15)

$$\frac{2\sqrt{(1-\lambda)r_1r_2} - y_n}{2\sqrt{(1-\lambda)r_1r_2} + y_n} = \frac{(\sqrt{r_2} - \sqrt{r_1})^n}{(\sqrt{r_2} + \sqrt{r_1})^n}.$$

Therefore

$$\sum_{n=1}^{\infty} \frac{2\sqrt{(1-\lambda)r_1r_2} - y_n}{2\sqrt{(1-\lambda)r_1r_2} + y_n} = \frac{\sqrt{r_2} - \sqrt{r_1}}{2\sqrt{r_1}}.$$

Also we have

$$\sum_{n=1}^{\infty} (y_n - y_{n-1}) = 2\sqrt{r_1(r_2 - r_1)}.$$

6. Some properties of conics associated with three mutually tangent circles. Let there be two circles (O_1) and (O_2) tangent internally at A (Fig. 4) and let the radii of these circles be r_1 and r_2 respectively, and let the radius of a circle tangent to these two and having its center on the line O_1O_2 be r_3 and let its center be O_3. Let O be the center of any circle tangent to (O_1) and (O_2) and let r be its radius.

The conic associated with (O) and (O_2) is an ellipse with the points O and O_2 as foci, and passing through O_1: the conic associated with (O) and (O_1) is a hyperbola passing through O_2 and having O and O_1 as foci. Likewise we will have an ellipse passing through O and having O_1 and O_2 as foci. Draw from A the line AT tangent to the circles (O_1) and (O_2). Then with the three circles $(O_1,)$ (O_2) and (O) the straight line AT there will be associated four parabolas[1] two of which pass through A, one through O_1 and the fourth through O_2.

[1] See article "Some Properties of a Straight Line and Circle and their Associated Parabolas," *Annals of Math.*, second series, Vol. 19, pp. 174–5. Also "Some properties of circles and related conics," *Annals of Math*, second series, vol. 20, pp. 279–280.

Call the conic through O_2, H_2, the one through O_1, E_1 and the one through O and O_3, E_3, and the parabolas through O_1 and O_2, P_1 and P_2 respectively.

With this notation the following may be readily proved analytically:

THEOREM: The normals to E_1 and H_2 at the points O_1 and O_2 intersect on a line through O_3 perpendicular to O_1O_2.

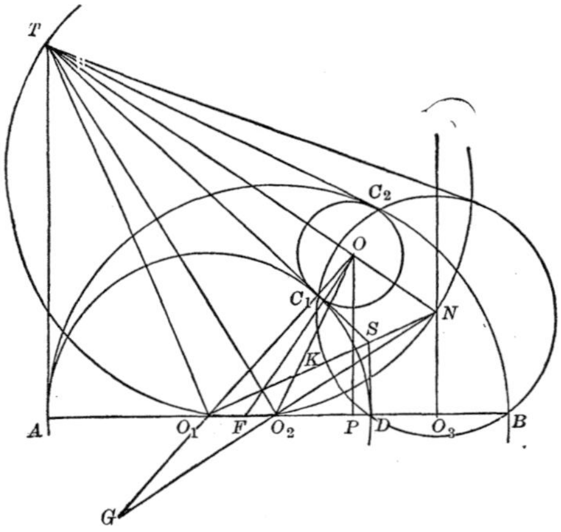

FIG. 4.

THEOREM: The tangents to H_2 and E_1 at the points O_2 and O_1 intersect on the line AT.

THEOREM: If two circles are tangent internally, and any circle is tangent to these two, there are associated with these circles three non-degenerate conics, six points of which are the three points of contact of the three circles and the three centers of the three circles, and the six tangents to the conics at these six points pass through a common point.

Proof: Let the normal to E_1 at O_1 and to H_2 at O_2 intersect at N, and let the intersection of the tangents at these points be T. Let the point of contact of (O) and (O_1) be C_1 and of (O) and (O_2) be C_2. Then the angle AO_1C_1 is bisected by O_1T. Therefore a line from T to C_1 will be tangent to (O_1) and (O). Moreover since T and N are ex-centers of the triangle OO_1O_2, T, N, and O are collinear. It is also evident that the tangent to (O_2) at C_2 will pass through T. We have therefore the six lines TA, TO, TO_1, TC_2 TO_2, and TC_1, and these lines are the six tangents to E_3, E_1 and H_2 at the points A, O, O_1, C_2, O_2, C_1.

THEOREM: E_1 and P_1 have the same normal at O_1.

Proof: Since O is the focus of P_1 and the axis is parallel to O_1O_2, then the bisector of the angle OO_1O_2 will be normal to P_1. But this is also normal to E_1 because O and O_2 are the foci of E_1.

THEOREM: The three axes of the three non-degenerate conics associated with three tangent circles, and the three normals at the centers of the circles, meet in points that are collinear.

Proof: Let the normal and axis of E_3 intersect in F, the normal and axis of H_2 intersect in G and the normal and axis of E_1 in K. Then since we have the triangle O_1O_2O and the two bisectors of two interior angles and the bisector of the opposite exterior angle, the points F, K and G are collinear.

The right angles formed by the tangents and normals at O_1 and O_2 are inscribed in a semicircle with TN as diameter. Call this circle C_t. Steiner has pointed out the fact that D, B, C_1 and C_2 are points of a circle C_n with center N.[1] It is then evident that the tangents to C_n at its points of intersection with C_t pass through T. We then have two sets of coaxial circles C_n and C_t, the centers

[1] See reference to Steiner in Introduction.

of one being on a line through O_3 perpendicular to O_1O_2 and the diameters of the other being segments of tangents to E_3 cut off by the tangents at A and O_3.[1] It should also be noted in this connection that the point T is the pole of the line drawn from the point of tangency of any two of the circles to the center of the third circle with respect to the conic passing through that center.

Also if there is drawn at D a line perpendicular to O_1O_2 and TC_1 is produced intersecting this line in S, then N, S and O_1 are collinear.

THEOREM: If three circles are mutually tangent and tangents and normals be drawn to the three associated conics at the points of contact and the centers of the three circles, and if the normals of two of the conics be chosen, these will intersect by twos on a tangent to the third conic.

Proof: Consider the lines O_1N, O_2N, OO_1, OO_2. These intersect in the points O and N which are on the tangent to E_3.

THEOREM: The axes and normals to two of the conics, together with the tangents to these two conics drawn at the centers of the circles determine two perspective triangles whose center of perspectivity is the intersection of the axis and normal to the third conic.

Proof: Consider the lines O_1N, O_2N, OO_1, OO_2, O_1T, O_2T. These intersect by twos on the line TN. They may therefore be considered as the sides of two perspective triangles. Let the corresponding sides be

$$\begin{array}{cc} O_1T, & O_2T, \\ O_2N, & O_1N, \\ OO_1, & OO_2. \end{array}$$

These determine the perspective triangles $A_{12}A_{13}A_{23}$ and $B_{12}B_{13}B_{23}$ and the center of perspectivity is the point F on O_1O_2 (see Fig. 4 where F is marked). But this point is also on the normal OB_{12}.

Corollary: There will be three such sets of perspective triangles and the centers of perspective will be collinear (second theorem before the last).

THEOREM: N and T are double points of an involution, of which O and the point (Y), where TN intersects O_1O_2, are a conjugate pair, and therefore O and Y are inverse points with respect to the circle C_t.

The proof of this theorem follows immediately from a consideration of the quadrangle $O_1A_{12}O_2B_{12}$.

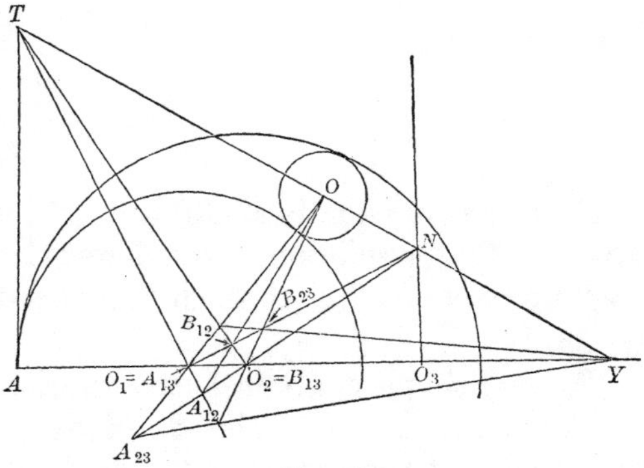

FIG. 5.

The following theorems are also evident.

THEOREM: If the three axes and the three normals to the three associated

[1] See *Conics* of Apollonius, Book III, prop. 45 (Heath's ed., p. 114).

conics be drawn the axis and normal to each conic being taken as corresponding sides, they form two perspective triangles with T as center of perspective.

THEOREM: The four axes of perspective of the three circles and the six lines, three of which are normals and the other three are the tangents to the three conics at the three centers of the three circles, determine a quadrangular-quadrilateral configuration, whose diagonal triangle is the triangle determined by the three centers of the three circles.

In connection with the above discussion it should be noted that it furnishes a method for constructing points on a conic. For let O_1, O_2 and O_3 be any three points on a line, and let the perpendicular be drawn at O_3 and let N be any point in the perpendicular. Let O_1 and O_2 be points such that we have the order $O_1O_2O_3$ or $O_2O_1O_3$. Then from N draw lines to O_1 and O_2, making the angles NO_1O_3 and NO_2O_3, and draw from O_1 and O_2 the lines O_1O and O_2O such that

$$\angle NO_1O_3 = \angle OO_1N,$$

$$\angle NO_2O_3 = \angle OO_2N.$$

Then the point O is on an ellipse. If we have the order $O_1O_3O_2$, O will be on a hyperbola, and if O_1 or O_2 is at infinity we have a parabola. And in each instance O_1 and O_2 are foci of the conic. This also gives a method for establishing a (1, 1) correspondence between the points of a conic and the points of a straight line.

7. Some Projective Properties of the Figure in Section 2. Let A and C be the ends of the diameter O_1O_2 of the circle (O_2) (Fig. 1), and let there be drawn from A lines to S_n and from C lines to S_n', and let C and C' be the angles that these lines make with AC. Then

$$\tan C = \frac{2nr_3}{r_1 + r_2}, \tag{16}$$

$$\tan C' = \frac{2nr_1}{r_2 + r_3}. \tag{17}$$

By means of equations (16) and (17) we may find the equations of the lines AS_n and CS_n', a solution of which reveals the fact that the line AS_n and the line CS_n' intersect on a line through S_1 perpendicular to AC in points whose ordinates are $2\rho_n$.[1]

By very simple analytical considerations we may prove the following

THEOREM: The triangles $S_{n+1}S_nS_{n-1}$ and $S_{n+1}'S_n'S_{n-1}'$ are perspective and their center of perspective is the external center of perspective of the circles O_1 and O_3.

THEOREM: The locus of the point of contact of two tangent circles which are tangent to two given tangent circles (internally tangent) is a circle whose center is the center of perspective of the two given circles and whose radius is the harmonic mean between the radii of the two given circles.

[1] In this connection see Steiner, p. 69 and ff.

Proof: The center of perspective, P_3, of the two given circles, O_1 and O_2, has the same power with respect to all circles tangent to these two in a given way. Therefore, the locus of the point of contact of any two such circles which are tangent to each other is a circle orthogonal to them all.

THEOREM: The circle with P_3 as center and P_3A as radius cuts every C_n orthogonally.

Proof: Let there be drawn with T as center and TA as radius a circle. This will pass through C_1 and C_2. Therefore $C_1C_2P_3$ will be the radical axis of the circle just drawn and C_n. The circle with P_3 as center and P_3A as radius is orthogonal to the circle with center T. It is therefore orthogonal to every C_n.

LIFE

I, James Henry Weaver, was born in Madison County, Ohio, June 19, 1883. I received my elementary education in the public schools of Ohio. I was granted the A.B. degree at Otterbein College in 1908 and the M.A. degree at Ohio State University in 1911.

I taught in the rural schools of Ohio during the years 1900–1904; was a Teacher and Principal in the High School at Plain City, Ohio, 1908–1910; was substitute instructor in the Department of Mathematics at Ohio State University, 1910–1912; and Head of Mathematics Department of West Chester High School, West Chester, Pa., 1912–1917.

I have published a number of short articles in the Annals of Mathematics, Bulletin of the American Mathematical Society, and School Science and Mathematics. The present paper was undertaken at the suggestion of Professor M. J. Babb, whose criticisms and suggestions have been most valuable.

[Intentionally Left Blank]

[Intentionally Left Blank]

[Intentionally Left Blank]

[Intentionally Left Blank]

[Intentionally Left Blank]

Made in the USA
Middletown, DE
05 November 2026

36560128R00015